Samuel Awadhifo Ayibho

Reciprocity and contrapositive in the Pythagorean theorem

Samuel Awadhifo Ayibho

Reciprocity and contrapositive in the Pythagorean theorem

ScienciaScripts

Imprint

Cover image: www.ingimage.com

This book is a translation from the original published under ISBN 978-620-6-69780-0.

Publisher:
Sciencia Scripts
is a trademark of
Dodo Books Indian Ocean Ltd. and OmniScriptum S.R.L publishing group

120 High Road, East Finchley, London, N2 9ED, United Kingdom
Str. Armeneasca 28/1, office 1, Chisinau MD-2012, Republic of Moldova, Europe
Printed at: see last page
ISBN: 978-620-7-89832-9

Contents

INTRODUCTION

In this work, we attempt to explain reciprocity and contraposity in the Pythagorean theorem[1] .

Traditionally, a theorem was explained as a structure made up of the following

[1] Pythagoras was a religious reformer and pre-Socratic philosopher who is thought to have been born around 580 BC on Samos, an island in the south-eastern Aegean Sea. His death is thought to have occurred around 495 BC , at the age of 85. He would have been also mathematician and scientist, according to a late tradition. The name Pythagoras (etymologically, *Pyth-agoras:* "he who was announced by the Pythia"), derives from the announcement of his birth made to his father during a trip to Delphi. The enigmatic life of Pythagoras makes it difficult to shed light on the story of this religious reformer, mathematician, philosopher and miracle-worker. It is well known that he never wrote anything, and the seventy-one lines of the *Golden Verses* attributed to him are apocryphal and are a sign of the immense development of the legend formed around his name. Neopythagoreanism is nevertheless imbued with a mysticism of numbers, already present in the thought of Pythagoras. Herodotus refers to him as "one of the greatest minds of Greece, the wise Pythagoras". He retained great prestige; Hegel said he was "the first universal teacher". According to a striking echo of Heraclides of Pontus, mentioned by Cicero, Pythagoras was the first Greek thinker to call himself a *philosophos*, meaning "friend of knowledge or wisdom". In *Tusculanes*, Cicero explains: "By the same reason, no doubt, all those who have since attached themselves to the contemplative sciences have been considered Wise Men, and have been called such, until the time of Pythagoras, who was the first to put the name of philosopher into vogue. Heraclides of Pontus, a disciple of Plato and a very clever man himself, tells the story. One day," he says, "Leo, king of the Phliasians, heard Pythagoras speak on certain points with such knowledge and eloquence that the prince, seized with admiration, asked him what art he was practising. Pythagoras replied that he knew none, but that he was a philosopher. The king, surprised at the novelty of this name, asked him to tell him who the philosophers were and how they differed from other men". Diogenes explains this about the transmigration of the soul: "He (Pythagoras) said the following about himself: he had once been Aithalides and was said to be the son of Hermes; Hermes had told him to choose whatever he wanted, except immortality. Hermes had told him to choose whatever he wanted, except immortality. He had therefore asked to keep the memory of what happened to him, both alive and dead. So in life he remembered everything, and in death he kept his memories intact. Later, he entered Euphorbia's body and was wounded by Menelaus. And Euphorbius said that he had been Aithalides [son of Hermes], and that he had inherited from Hermes this gift and the way in which the soul passed from one place to another, and he told how it had accomplished its journeys, in which plants and animals it had found itself present, and all that his soul had experienced in Hades, and what the others endured there. When Euphorbius died, his soul passed into Hermotime who, wanting to give proof himself, returned to the Branchidae and, entering the sanctuary of Apollo, showed the shield that Menelaus had consecrated there (he said in fact that Menelaus, when he set sail from Troy, had consecrated this shield to Apollo), a shield that had decomposed by that time, of which only the ivory face remained. When Hermotime died, he became Pyrrhos, the delirious sinner; again, he remembered everything, how he had previously been Aithalides, then Euphorbia, then Hermotime, then Pyrrhos. When Pyrrhos died, he became Pythagoras and remembered everything that has just been said. (Source: Wikipedia. Biography of Pythagoras, available on Pythagoras - Wikipedia (wikipedia.org) Accessed Thursday 2 November 2023 at 10:37.

elements[2] :

- Assumptions: i.e. basic conditions that are listed in the theorem in addition to the elements already presented in the theory;
- a thesis, also called a conclusion: i.e. a mathematical or logical statement that the theorem proves to be true under the basic assumptions;
- the demonstration: as a theorem can sometimes be demonstrated in several very different ways (see the example of the multiple demonstrations of the Pythagorean theorem), only the fact that the demonstration exists is a constituent of the theorem, but not the details of the demonstration. A demonstration is a sequence of logical inferences involving the axioms of the underlying theory, the hypotheses of the theorem, and results previously established within the framework of that theory.

This raises the question:

- What can we learn from Pythagoras' theorem?
- What is reciprocity?
- What about the contrapositive?

Providing answers to these questions will be our task in this work.

In this work, we formulate the hypotheses according to which :

- The generality of the Pythagorean theorem would be a real key to understanding reciprocity and contraposing.
- Reciprocity would be an essential point in the Pythagorean theorem
- It would be wrong to talk about the Pythagorean theorem without explaining the principle of the contrapositive.

The general aim of this work is to understand the Pythagorean theorem, explaining its reciprocity and contrapositive. In the same perspective we have three specific objectives, exactly to present:

- the generality of the theorem,
- reciprocity

[2] Wikipedia. Theorem, available on Theorem - Wikipedia (wikipedia.org) Accessed Thursday 2 November 2023 at 10 :48.

- the contrapositive.

We used an analytical method, studying and documenting the Pythagorean theorem from the perspective of reciprocity and contraposity. The technique used is documentary. We used documents and data on the pages consulted.

This theme has been chosen from a number of perspectives: philosophical, geometric, mathematical, etc.

We have three axes: the first deals with the generality of the Pythagorean theorem, the second explains reciprocity and the last explains the contrapositive.

1 THEORY

1.1 Theorems in general

Theorems are used in many fields, including mathematics, physics and philosophy. The word theorem is derived from the Greek word *theorein*, which translates as "to look at" or "to contemplate". From the same perspective, we can note[3] :

- In mathematics, a theorem is an assertion that can be demonstrated on the basis of certain hypotheses. The demonstration of a mathematical theorem is a chain of logical arguments that lead inexorably to the conclusion of the theorem.
- In physics, a theorem is a law that describes a physical phenomenon precisely. Theorems in physics are generally stated in the form of mathematical laws. The law of universal gravitation, for example, is a theorem of physics.
- In philosophy, a theorem is a proposition that can be explained on the basis of certain hypotheses. The demonstration of a philosophical theorem is a chain of logical arguments that lead inexorably to the conclusion of the theorem. This is why we have the Pythagorean theorem, which everyone does at school.

For example, we have Thales' theorem[4] . The theorem of Thales, my dear budding geometers, is a little nugget of knowledge that will make your eyes and curious minds sparkle[5] ! You know, that magical theorem that establishes a spellbinding link between the sides of similar triangles, allowing you to solve proportion riddles like real number wizards. Well, the Thales Theorem exercises are a bit like a learning potion that transports you into the spellbinding world of

---

[3] Wikipedia. Definition of a theorem, available on Théorème : définition, applications et exemples - Progresser-en-maths Accessed Thursday 2 November 2023 at 11 :00.

[4] Thales, son of Examyes and Cleobulin, was born around 625 BC in the Greek city-state of Milet, in Ionia, now Turkey. He died in 547 B.C. His family was of Phoenician origin. According to the account given by the historian Herodotus, Thales of Miletus took an active part in the politics of his city, giving advice to the Ionians. He is described not only as a politician but also as a great engineer, having succeeded in digging a deep channel in the river Halys, allowing the waters to change direction, which enabled Croesus' army to cross to the other bank.(Wikipedia. Biography of Thales, available on Thales of Miletus: biography and philosophy (filosofiadoinicio.com) Accessed Thursday 2 November 2023 at 11 :29).

[5] Wikipedia. 18 worksheets on the theorem of Thales of Milet, available on 18 worksheets on the theorem of Thales - Prof Innovant Accessed Thursday 2 November 2023 at 11 :26.

geometry. So put on your mathematician's robes and get ready to plunge into this fascinating geometric adventure.

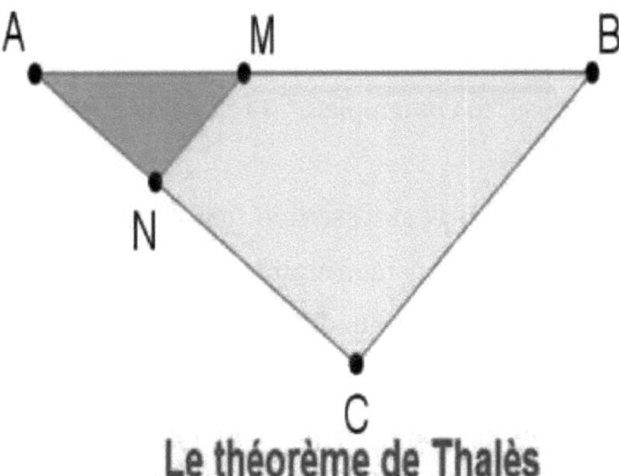

Le théorème de Thalès

Logical links between theorem, reciprocal and contrapositive[6] .

if the theorem is true	The reverse is not necessarily true	The converse is true
If the converse is true	The theorem is not necessarily true	The opposite is not necessarily true
If the converse is true	The theorem is true	The theorem is not necessarily true

Examples:

	A theorem with a true reciprocal.	**A theorem with a false reciprocal.**
Theorem	Multiples of two are even numbers.	Multiples of four are even numbers.
Reciprocal	Even numbers are multiples of two.	Even numbers are multiples of four. **False**, because 6 is not a multiple of 4!
Contraposed	If the number is not even,	If the number is not even, then it is

	then it is not a multiple of two.	not a multiple of four.

1.2 The Pythagorean theorem

The Pythagorean theorem[6] [7] is a theorem of Euclidean geometry that states the relationship between the lengths of the sides of a right-angled triangle. It is often stated in the following form:

If a triangle is right-angled, the square of the length of the hypotenuse (or side opposite the right angle) is equal to the sum of the squares of the lengths of the other two sides.

This theorem also allows one of the lengths to be calculated from the other two.

It owes its name to Pythagoras of Samos, an ancient Greek philosopher of the VI[e] century BC, but the result was known over a thousand years earlier in Mesopotamia and was probably discovered independently in several other cultures.

The earliest known explanation comes from Euclid, around 300 BC, and although Greek mathematicians also knew one before that, there is no way of attributing it with any certainty to Pythagoras.

The first historical demonstrations were generally based on methods of calculating area by cutting out and moving geometric figures. Conversely, the modern conception of Euclidean geometry is based on a notion of distance that is defined to respect this theorem.

Various other statements generalise the theorem to any triangles, to higher-dimensional figures such as tetrahedra, or in non-Euclidean geometry to the surface of a sphere.

More generally, this theorem has many applications in very different fields

---

6 Wikipedia; logical link between theorem , available at What is the difference between theorem, reciprocal and contrapositive - Objective: succeed in maths (objectif-reussir-en-maths.com) Accessed on Thursday 2 November 2023 at 15:48.

7 Wikipedia. Theorem, available on Pythagorean theorem - Wikipedia (wikipedia.org) Accessed Tuesday 31 October 2023 at 9 :7.

(architecture, engineering, etc.), even today, and has led to many technological advances throughout history.

Exercises on the Pythagorean theorem[8]

Exercise 1

The triangle XYZ is such that XY = 29.8 cm, YZ = 28.1 cm and XZ = 10.2 cm. Explain why it is not a right-angled triangle.

[8] Wikipedia. Exercises on the Pythagorean theorem, available on Pythagorean theorem, course + corrected exercises. (paramaths.fr) Accessed Thursday 2 November 2023 at 11:15.

Exercise 2

Calculate the value, rounded to the nearest millimetre, of :

a. the length of the diagonal of a square with sides of 5 cm

b. the length of the diagonal of a rectangle whose dimensions are 8.6 cm and 5.3 cm

c. the length of the side of a square with a diagonal of 100 m.

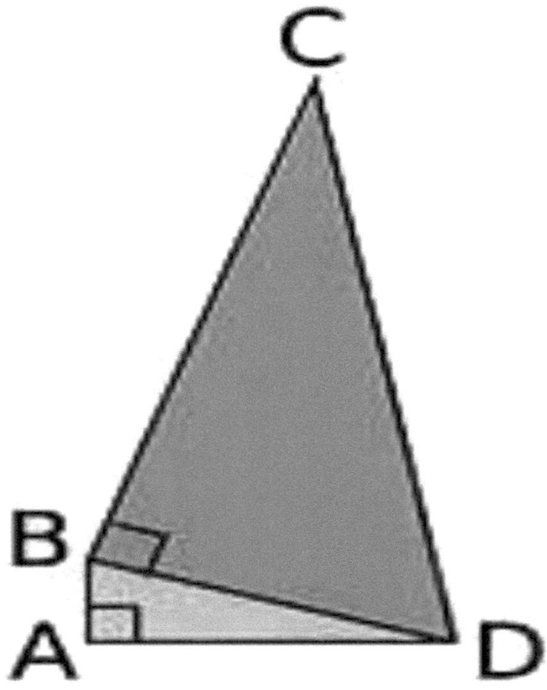

In the figure above: AB = 1.5 cm; AD = 6 cm and BC = 12 cm

1. Calculate the value, rounded to the nearest mm, of BD

2. Calculate and justify the exact value of DC

Exercise 4

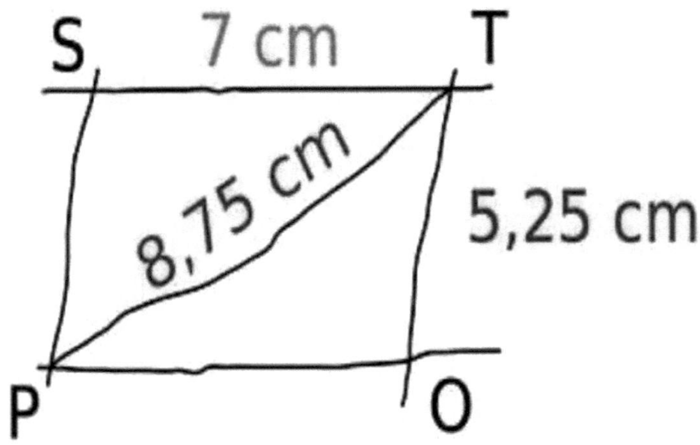
S
7 cm
T
8,75 cm
5,25 cm
P
O

2 THEOREM RECIPROCITY

2.1 Reciprocity in general

Mathematically, in propositional calculus, a reciprocal implication is a proposition that exchanges the premise and conclusion of an implication. The reciprocal of the reciprocal then reverts to the initial implication. When the implication has several premises, the exchange of the conclusion with only some of the premises is sometimes also called a reciprocal, as in the case of Thales' theorem, where the alignment conditions remain in the premise for the reciprocal[9] . Unlike the contrapositive of an implication, the reciprocal cannot be deduced from the implication. To do so without caution leads to the fallacy of asserting the consequent, so let's take this example[10] :

The implication "if A then B" has as its reciprocal, "if B then A".

or .

This notion of reciprocal implication is sometimes extended to the calculus of predicates by saying that either "all A is B" or "all B is A" are reciprocal implications of each other.

However, a sentence of the form "no A is B" is equivalent to "no B is A". Their common reciprocal can be stated as "everything that is not A is B".

Truth table of an implication and its reciprocal

P	Q	$P \rightarrow Q$	$Q \rightarrow P$ (reciprocal)
V	V	V	V
V	F	F	V
F	V	V	F
F	F	V	V

2.2 Pythagorean reciprocity

We can see that in addition to the Pythagorean theorem, there is its reciprocal,

[9] Wikipedia. Reciprocity, available on Reciprocal involvement - Wikipedia (wikipedia.org) Accessed Thursday 2 November 2023 at 13 :59.

[10] *Ibid.*

which means that we are going to go into the other meaning: we are going to show that a triangle is right-angled from the measures of the sides of a triangle[11]

:

- the reciprocal of the Pythagorean theorem is as follows:

In a triangle, if the square of one side is equal to the sum of the squares of the other 2 sides, then the triangle is right-angled.

- Let's explain this reciprocal with an example: let's take the EFG triangle

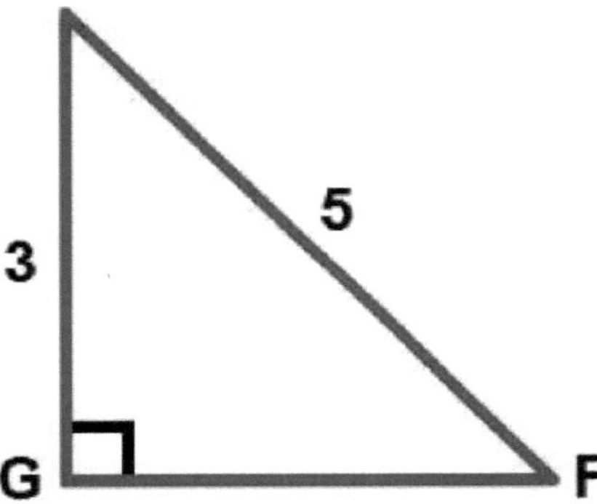

next :

Let's calculate EF^2 separately, and $EG^2 + GF^2$ separately. $EF^2 = 5^2 = 25$ and $EG^2 + GF^2 = 3^2 + 4^2 = 9+16 = 25$ So $EF^2 = EG^2 + GF^2$

Here EF^2 represents "the square of one side", and $EG^2 + GF^2$ represents "the sum of the squares of the other 2 sides". And from this point of view, we've just explained that it was "equal".

So, according to the reciprocal of the Pythagorean theorem, the triangle EFG is a right-angled triangle. As you can see, it's nothing too serious, but we're going to come back to some important points.

First of all, the choice of sides: why was EF^2 calculated, and not EG^2 or FG ?^{2}

Quite simply because it is the largest of the 3 sides, as we saw in the Pythagorean theorem that the "only" side was the hypotenuse, and therefore the

[11] Wikipedia. Pythagorean theorem, available on Le théorème de Pythagore | Méthode Maths (methodemaths.fr) Accessed Monday 30 October 2023 at 16 : 13

largest.
The squared side that is calculated "on its own" is therefore the larger side.
Next, it is very important to do both calculations SEPARATELY!
In the example[12] [13] we want to show that $EF^2 = EG^2 + GF^2$, this is the conclusion, so we don't start from this equality because we don't know if it's true, we want to show it.
We therefore calculate first EF^2 then $EG^2 + GF^2$.
There's nothing to stop you doing the calculations at the same level, which is what we'll do in the next example.
Note: the conclusion is that the triangle is right-angled. Yes, but at what point? Well, in the example, EF is the longest side, i.e. the hypotenuse, so the right angle is opposite, at G.
So there's nothing to stop you saying in the conclusion "so the triangle EFG is right-angled at G", in fact it's much better to say so!
Let's look at an example using the wording: let's show that the following triangle PQR is right-angled:

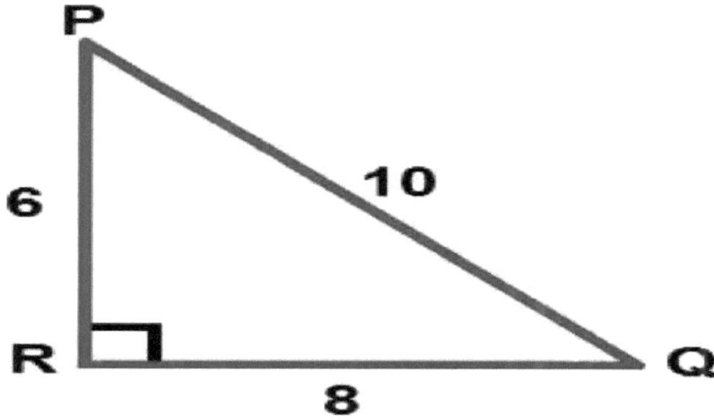

$PQ^2 = 10^2$ and $PR^2 + RQ^2 = 6^2 + 8^2$

$PQ^2 = 100$ and $PR^2 + RQ^2 = 36 + 64$

$PQ^2 = 100$ and $PR^2 + RQ^2 = 100$

12

13 Wikipedia. Pythagorean theorem, available on Le théorème de Pythagore | Méthode Maths (methodemaths.fr) Accessed Monday 30 October 2023 at 16 : 13.

13

So $PQ^2 = PR^2 + RQ^2$

So, according to the reciprocal of the Pythagorean theorem, triangle PQR is right-angled at R.

As you can see, it's quite quick after all!

Note[14] : in this case, the 2 calculations have been performed side by side (PQ^2 and $PR^2 + RQ$).^{2}

separated by ""and"". And it was only when we finished the calculation and arrived at the same result that we said it was equal.

However, sometimes at the end of the calculation you don't find the same thing, so it's not equal and you can't apply the theorem, and the triangle isn't right-angled!

That's why we have the contrapositive, which we're going to look at more closely.

The reciprocal of the Pythagorean theorem is used to explain that a triangle is right-angled.

Consider a triangle whose lengths are 3, 4 and 5. Can you work out that this is a right-angled triangle[15] ?

To explain that this triangle is a right-angled triangle, we need to use the reciprocal of the Pythagorean theorem.

On the one hand, we have 32+42=9+16=25.

On the other hand, it is true that 52=25.

Then 32+42=52, and by the reciprocal of the Pythagorean theorem, this is a right-angled triangle.

Knowing that the hypotenuse is the longest side in a right-angled triangle. So, if we need to explain that a triangle is a right-angled triangle, we need to sum the squares of the two shortest lengths and check that this sum is equal to the square

---

[14] Wikipedia. Pythagorean theorem, available on Le théorème de Pythagore | Méthode Maths (methodemaths.fr) Accessed Monday 30 October 2023 at 16 : 13

[15] Wikipedia. Pythagorean theorem formula, available on Pythagorean Theorem : lessons and exercises | StudySmarter Accessed Thursday 2 November 2023 at 9 :45.

of the longest side[16] .

For a better understanding of the theorem, let's look at[17] :

Pythagorean theorem - Key points

- The Pythagorean theorem is used to solve problems involving a right angle or a right-angled triangle.
- The Pythagorean theorem states that the square of the hypotenuse of a right-angled triangle is equal to the sum of the squares of the other two sides: a +b =c[222] .
- We can use the reciprocal of the Pythagorean theorem to show that a triangle is right-angled.
- Similarly, we can use its contrapositive to check that a triangle is not right-angled.

To calculate the height of a mountain or a building to be constructed, we can use the Pythagorean theorem. There are many applications for this theorem, and it is often regarded as one of the most important theorems in mathematics[18 19] .

The Pythagorean theorem is mainly used to solve problems involving right-angled triangles. Among others, we can use the Pythagorean theorem for 18:

- determine unknown lengths ;
- calculate minimum distances between two objects ;
- calculate the forces acting on a structure ;
- derive other theorems.

The Pythagorean theorem has applications in many fields, such as engineering and computing. Bear in mind that for 'real life' applications, the Pythagorean theorem is often used in conjunction with other more advanced theorems.

---

[16] *Ibid.*

[17] *Ibid.*

[18] Wikipedia. Pythagorean theorem formula, available on Pythagorean Theorem : lessons and exercises | StudySmarter Accessed Thursday 2 November 2023 at 9 :45.

[19] *Ibid.*

3 THEOREM CONTRAPOSED

3.1 General information on the contrapositive

In logic[20] , contraposition is a type of reasoning used to assert the implication "if not B then not A" from the implication "if A then B ". The implication ""if not B then not A" " is
called the contrapositive of "if A then B".

For example, the contrapositive of the proposition ""*if it rains, then the ground is wet*"" is ""*if the ground is not wet, then it does not rain*"".

For example, the contrapositive of the proposition "*if n is a multiple of 9, then n is not prime*" is "*if n is prime, then n is not a multiple of 9*".

This is why Aristotle says "In order to bring things back to the principles of this theory, we make up the lengths of the long and the short, that is to say of a certain species of large and small, the surface of the wide and the narrow, the body of the deep and its opposite"[21] .

3.2 The Pythagorean contrapositive

As we know, the contrapositive of the Pythagorean theorem is used to explain that a triangle is not right-angled if we know the lengths of the three sides of a triangle[22] .

In a triangle, if the square of the longest side is equal to the sum of the squares of the other two sides, then the triangle is right-angled.

---

[20] Wikipedia. La contraposée, available at contraposée | Lexique de mathématique (netmath.ca) Accessed Thursday, November 2, 2023 at 12:8.

[21] ARISTOTE, La métaphysique, Paris: Ladrange, (1838), (1840), 2008, p.35-36.

[22] Wikipedia. Contrapositive of Pythagorean theorem, available on Révision quatrième du théorème de Pythagore (free.fr) Accessed Tuesday 31 October 2023 at 8 :43.

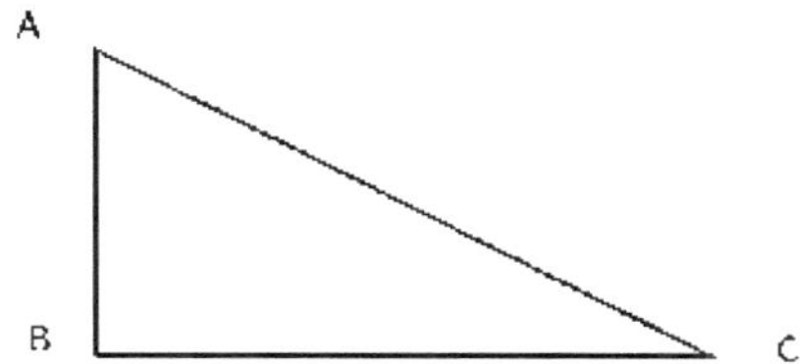

If $AC^2 = AB + BC^{22}$,

Then ABC is a right-angled triangle.

Exercise[22] .

Exercise 1: In each case, calculate the length BC:

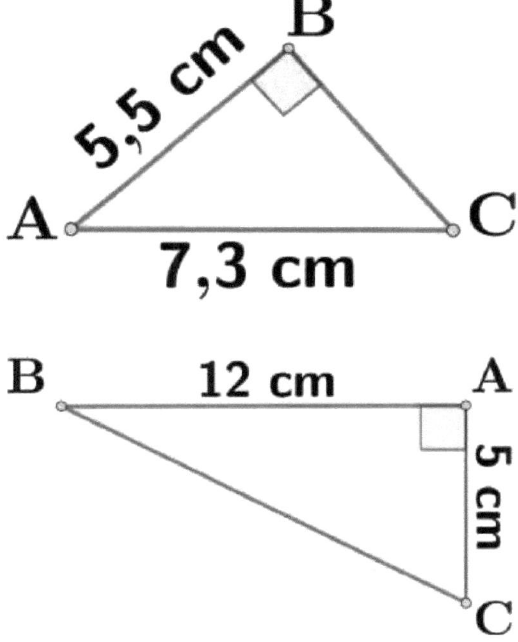

Find out if a triangle is right-angled using d

Specify whether the triangles BOA and NEZ are rectangles. Justify your answer.

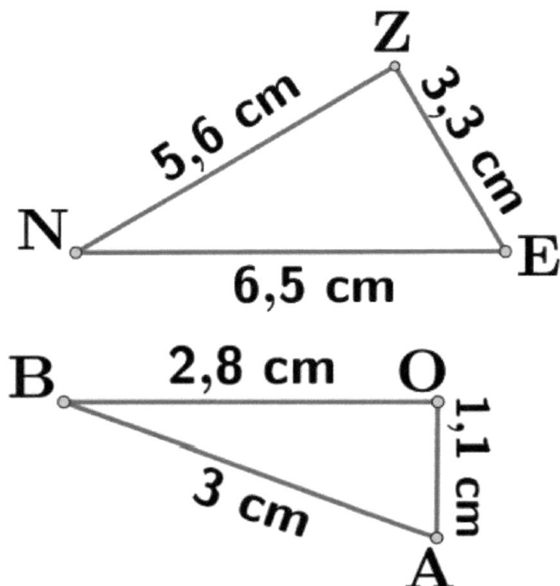

Calculate the length AN to the nearest tenth of a centimetre:

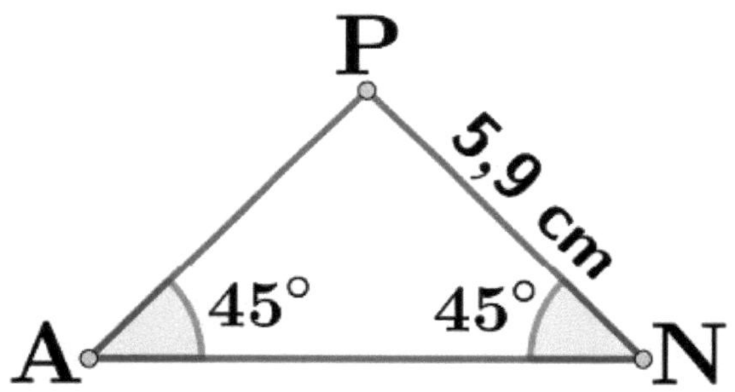

A triangle ABC has been drawn on a square grid. Is this triangle right-angled?

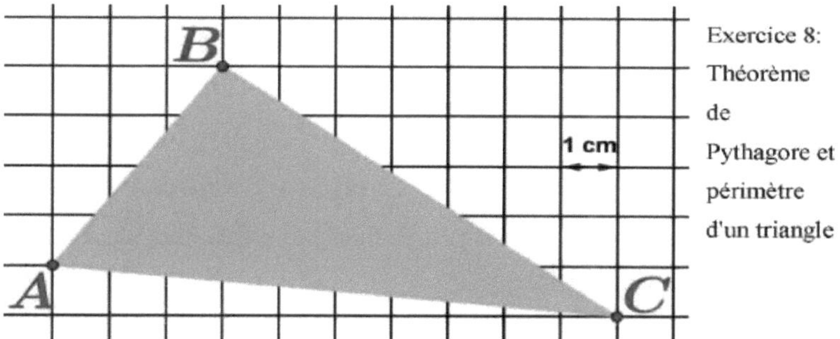

Let x *be* a positive number. Consider a triangle whose sides measure $3x+1$, $4x+3$ and $5x+3$. Is this triangle right-angled?

CONCLUSION

We have completed, but not finished, the explanation of reciprocity and contrapositive in the Pythagorean theorem. There were three main strands to our work.

First, we presented the theorem. In the first place, we were concerned with the generality of the notion of theorem, and then we explained the Pythagorean approach.

Secondly, there was the understanding of reciprocity. After trying to determine the general meaning, we explained this reciprocity in the Pythagorean theorem.

Thirdly, the contrapositive. We explained the notion of reciprocity in the general sense, and then in the Pythagorean theory.

An ideal is created on a triangle if it is a right-angled triangle. It is from this perspective that the idea of reciprocity and contraposity appears in the Pythagorean theorem. This is the possibility of taking several forms to an art.

As we know: "The ideal cannot remain a mere abstract concept. By virtue of its very idea, it contains a specific and particular element. It must therefore manifest itself in a specific form. The question then arises as to how the ideal, while passing into the external and finite world, retains its own nature, and how the latter, for its part, becomes capable of receiving into its bosom the ideal principle that constitutes art"[23] .

---

[23] G.W.F. Hegel, Esthétique, Quebec, 1835, p.70.

List of theorems in alphabetical order, A to Z [24]:

A

- Abel's theorems
- AF+BG theorem
- Arc theorem
- Abelian and Tauberian theorems
- The theorem of finite increments
- Acting theorem
- Ado's theory
- Alasia Theorem
- Fundamental theorem of algebra or d'Alembert-Gauss theorem
- Theorem Alexandrov
- Al-Kashi Theorem
- Alternative theorems
- Ampère's theorem
- Fundamental theorem of analysis
- The inscribed angle and centre angle theorem
- Cancellation theorem
- Apéry's theorem
- Open application theorem
- Learning theorem
- Universal approximation theorem
- The capable arc theorem
- Archimedes' theorem
- Argand's theorem

[24] Wikipedia. List of theorems, available on List of theorems - Wikipedia (wikipedia.org) Accessed Thursday 2 November 2023 at 16 :15.

- The argument theorem
- Fundamental theorem of arithmetic
- Arrow's impossibility theorem

- Artin's theory
- Artin-Schreier theory
- Artin-Whaples theorem
- Arzelà-Ascoli theorem
- Atiyah-Singer index theory
- TheoremATS
- Aubry's theory
- Ax-Katz Theorem
- Ax-Kochen theorem

B

- Theorem Babinet
- Bachet-Bézout theorem
- Theorem Baire
- Baire 's simple limit theorem
- Theorem Baire-Banach
- Baker theorem
- Theorem Balian-Low
- Theorems of Banach *et al.*
- Bartle-Graves theorem
- Finite base theory **(of)**
- Complete base theorem
- Normalbasis theory
- Bayes theorem
- Theorem Beatty
- Bell Theorem
- Bernoulli Theorem
- Bernstein's theorems
- ThéorèmedeBertrand
- Theorem Beth
- Theorem Beurling-Lax

- Bézouttheorem
- Bieberbach's conjecture, demonstrated by de Branges
- Theorem Bieberbach
- Wellness theory
- Bijection theorem
- Theorem Bing-Nagata-Smirnov
- Theorem Birch
- Birkhoff theorems
- Bisector theory
- Theorem deBloch
- Theorem Blumberg
- Theorem Bochner
- Theorem Bochner-Weil
- Theorem Bohr-Mollerup
- Theorem Bolillier
- Theorem Bolzano-Weierstrass
- Theorem Bombieri-Vinogradov
- Bonnet theorems
- Bonnet-Schoenberg-Myers theorem
- Borel's law of zero one
- Borel's theorem
- Borel-Cantelli theorem
- Borel-Lebesgue theorem
- Boundary theorem
- Upper bound theorem
- Borsuk-Ulam theorem
- Bose-Parker-Shrikhande theorem
- Boucherot theorem
- Hairy ball theorem
- Brahmagupta's theorem

- Brauer theorems
- Brauer-Siegel theorem
- Breusch theorem
- Brianchon theorem
- Brianchon-Poncelet theorem
- Brooks theorem
- Brouwer domain invariance theorem
- Brouwer fixed point theorem
- Browder's fixed point theorem
- Browder's selection theorem
- Brun's theorem
- Budan theorem
- Burke's theorem **(The Burke Theorem)**
- Burnside theorems

C

- Calderón's theorem
- Cameron-Martin theorem **(Theorem)**
- Cantor's theorem
- Cantor-Schroder-Bemstem theorem
- Carathéodory's theorem (geometry)
- Carathéodory extension theorem
- Caristi fixed point theorem
- Carnot's theorems
- Square theorem
- Four-square theorem
- Chetaev's theorem
- Theorems of Cartan Theorems
- Cartan-DieudonnéTheorem
- Theorem Cartan-Hadamard
- Cartan-von Neumann theorem

- Theorem Castigliano
- Catalan Theorem
- Cauchy theorems
- Cauchy-Peano-Arzelà theorem
- Theorem Cauchy-Kowalevski
- Theorem Cauchy-Lipschitz
- Cavalieri Theorem
- Cayley Theorem
- Theorem Cayley-Hamilton
- Theorem Cayley-Salmon
- Six-circle theorem
- Cesàro's theorems
- Ceva Theorem
- Change of variable theory
- Charkovski Theorem
- Chasles theorem
- Chebotarev's density theorem
- Chen's theorem
- Chevalley-Waming theorem
- Choi's theorem
- Chomsky-Schützenberger theorems
- Choquet-Deny theorem
- Chowla-Mordell theorem
- Chudnovsky's theorem
- Church's theorem
- Clairaut's theorems
- Clapeyron's theorem **(in)**
- Coase theorem
- Spider web theorem
- Cobham's theorem

- Cohen's theorem
- Cohn's irreducibility criterion
- Theorèmeducollage
- Compatibility theorem
- Nested compact theorem
- Comparison theorems
- Completeness theorem for the calculation of propositions
- Linear fungruence theorem
- Theorem Connelly
- Dominant convergence theorem
- Monotonic convergence theorem, or Beppo Levi's theorem
- Conway-Norton conjecture, demonstrated by Borcherds
- Cook-Levin theorem
- Correspondence theorem
- Odds Theorem
- Four-colour theory
- Scissorstrike theory
- Cox Theorem
- Cox-Jaynes Theorem
- Theorem CPT
- Craig's interpolation theorem D
- Dandelin-Quételet theorem
- Darboux theorems
- Davenport-Cassels theorem
- De Bruijn-Erdos theorems
- Decomposition theorems
- Theorem Dedekind
- Theorem of reduction
- ThéorèmedeDejean
- ThéorèmedeDelsarte-McEliece

- De Morgan's laws
- Theorem Denjoy-Young-Saks
- ThéorèmedeDesargues
- Descartes theorem (algebra)
- Descartes theorem (geometry)
- Theorem Descartes-Euler
- Theorem Diaconescu
- Theorem Dilworth
- Dini theorems
- Dini-Lipschitz theorem **(Theorem)**
- Dirichlet theorems (arithmetic progression, Fourier series, units)
- Euclidean division theorem
- Dixon's theorem (or identity)
- Theorem of Dold-Thom
- Theorem of Don Chakerian
- Theorem of Donsker Theorem
- Doob's stopping theorem
- Duality theorems
- Dugundji extension theorem
- Duhamel's theorem
- Theorem of Duhem's theorem
- Dunford-Schwartz Dunford-Schwartz theorem
- deDupuis theorem
- Theorem of Dvoretzky Theorem
- Theorem of Dvoretzky-Rogers
- Devon Dyck theorem
- Dynkin's theorems

E

- Earnshaw's theorem
- Easton's theorem

- Eberlein-Smulian theorem
- Edelstein's theorem
- Egoroff's theorem
- Egorychev's theorem
- Egregium theorem
- Ehrenfest theorem
- Ehresmann's theorem
- Eilenberg-Zilber theorem
- The primitive element theorem
- Cut elimination theorem
- Elitzur's theorem
- Engel's theorem
- Equipartition theorem
- Erdos theorems
- Ergodic theorem
- Euclid's prime number theorem
- Eudoxis' theorems
- Euler's theorems
- Excision theorem
- Six exponentials theorem

F

- The invariant factor theorem
- Fagin's theorem
- Fagnano's theorem
- Faltings' theorem
- Fary-Milnor theorem
- Fatou's theorem
- Favard's theorem
- Fedorov's theorem **(Theorem)**
- Feit-Thompson theorem

- Fej ér's theorem
- Fenchel theorem
- Fermât theorems
- Theorem of nested closures (and nested segments)
- Ferrero-Washington Theorem
- Feuerbach Theorem
- Finiteness theorems
- Floquet Theorem
- Theoremflot-max/min-cut
- Fluctuation-dissipation theorem
- Folkman's theory
- Theorem of implicit functions
- Theoremof implicit functions holomorphic
- Fundamental theorems
- Vital forces theorem
- Theorem Fourier
- Theoremofréchet-Riesz
- Frechet-Kolmogorov Theorem
- Fredholm theorems
- Theorem Frégier
- Theorem Freiman
- Theorem Freudenthal
- Theorem Friedlander-Iwaniec
- Frobenius theorems
- Theorem Froda
- Theorem Fubini-Tonelli
- Fubini 's differentiation theorem
- Theorem Fuchs **(and)**
- Fujimoto-Green theorem

G

- Art gallery problem
- Galois theorem
- Gauss theorems
- Gauss-Bonnet theorem
- Gauss-Lucas theorem
- Gauss-Markov theorem
- Gauss-Wantzel theorem
- The frost theorem
- Gelfand-Mazur theorem
- Gelfond-Schneider theorem
- Fundamental theorem of affine geometry
- Fundamental theorem of projective geometry
- Theorem of the gendarmes, or the frame, or the sandwich
- Gentzen's theorem
- Gergonne theorem
- Theorem by Sophie Germain
- Gershgorin's theorem
- Gibbard-Satterthwaite theorem
- Gibbs' theorem
- Girard's theorem
- Girsanov's theorem
- Glaeser's theorem
- Gleason-Montgomery-Zippin theorem
- Glivenko-Cantelli theorem
- Gödel's completeness theorem
- Gödel's incompleteness theorems
- Goldstine theorem
- Goldstone theorem
- Golod-Chafarevitch theorem
- Goodstein's theorem

- Gordan's theorem
- Gougu's theorem
- Goursat theorem
- Drop theorem
- Gradient theorem
- Closed graph theorem
- Perfect graph theorems
- Graves-Lusternik theorem
- Grebe's theorem
- Green's theorem
- Green-Ostrogradski flow-divergence theorem
- Green-Tao theorem
- Gromov's theorem on polynomial growth groups
- Gromov's non-plongement theorem
- Theorem Gronwall on the divisor function
- Theorem Grothendieck-Riemann-Roch
- TheoremedeGua
- ThéorèmedeGuilbaud
- Guldin theorems

H

- Haar's theorem
- Haag's theorem **(The Haag Theorem)**
- Hadamard's three straight lines theorem
- Hadamard's three-circle theorem
- ThéorèmedeHadamard-Lévy
- Theorem Hahn-Banach
- Theorem Hahn-Jordan **(and)**
- Theorem Hahn-Mazurkiewicz
- Hall Theorem
- Theorem Hamilton

- Hardy Theorem
- Hardy-Littlewood Metauberian Theorem
- Theorem Hardy-Littlewood
- Harnack principle
- Theorem Hartman-Grobman
- Theorem Haruki
- Hasse theorems
- Theorem Hasse-Minkowski
- Development theory of Heaviside
- Heawood theorem
- Heckscher-Ohlin-Samuelson theorem
- Heine Theorem
- Theorem Heine-Borel
- Theorem Helly
- Theorem Herbrand
- ThéorèmedeHerbrand-Ribet
- Hermite-Lindemann theorem
- Theorem Hessenberg
- Theorem 90 Hilbert's
- Hilbert basis theorem
- Hilbert's irreducibility theorem
- Hilbert's syzygy theorem
- Hilbert zeros theorem
- Hilbert-Samuel Theorem
- Hilbert-Schmidt Theorem **(Theorem)**
- TheoremedeHilbert-Speiser
- Theorem Hilbert-Waring
- Theorem Hille-Yosida
- Theorem Hjelmslev **(and)**
- Hodge index theorem **(of)**

- Theorem Hohenberg-Kohn
- Theorem Hölder
- Holmgren's uniqueness theorem **(in)**
- Theorem Holz
- Theorem Hopf-Rinow
- Hospital Rule
- Theorem Hu
- Oiler theory
- Hurewicz theorem
- Hurwitz theorems
- Huygens theorem

I

- Principal ideal theorem
- Identity theorem
- Open image theorem
- Ingham's theorem
- Limit inversion theorem
- Series-integral inversion theorem
- Local inversion theorem
- Ionescu-Tulcea-Kolmogorov theorem
- Isomorphism theorems
- Theorem isoperimetric
- Iteration theorem
- Iwaniec-Richert theorem
- Iwasawa theory

J

- Jackson's theory **(and)**
- Theorem Jacobi
- Theorem Jacobi-Kronecker
- Theorem Jacobi-Liouville

- James's theory
- Japanese theorem
- Jeffrey-Kirwan theorem
- Jensen's theorems
- Joachimsthal theorems
- John 's theory **(and)**
- Johnson theorem
- Jones Theorem
- Theorem Jordan-Brouwer
- Theorem Jordan-Hölder
- Theorem Jordan-Schur
- Jung's theory
- Kakutani fixed point theorem
- Theorem Kantorovitch **(and)**
- Theorem Kantorovitch-Rubinstein
- Theorem Karamata
- Kelvin's flow theorem
- Theorem Kennelly
- Theorem Kirchberger
- Theorem Kirszbraun **(and)**
- Theorem Kleene
- Kleene 's fixed point theorem
- Kleene 's theorem of recursion
- Theorem Knaster-Tarski
- Theorem Kneser (combinatorics)
- Koebe quarter theorem
- Theorem Koksma
- Kolmogorov's law of zero one
- Kolmogorov extension theorem
- Three Kolmogorov series theorem

- Kolmogorov-Arnold-Moser theorem
- König theorems
- Theorem Korovkin
- Theoremede Krasnosel'skii
- Theorem Krein-Milman
- Theorem Kreisel-Friedman
- Theorem Kronecker (diophantine approximation)
- Theorem Kronecker-Schering
- Theorem Kronecker-Weber
- Krull's theorem
- Krull principal ideal theorem
- Krull intersection theorem
- Theorem Krull-Akizuki
- Theorem Krull-Schmidt
- Theorem Kruskal
- Theorem Kruskal-Katona
- Theorem Kummer
- Theorem Künneth
- Theorem Kuratowski
- Kuratowski 's closing/complementary theorem
- Theorem Kürschák
- Theorem Kutta-Jukowski

L

- Lagrange theorems
- Theorem Laguerre
- La Hire Theorem
- Theorem Laman
- Lambert Theorem
- Landau Theorem
- Theorem Lasker-Noether

- Theorem Lasota-Yorke
- Theorem Lax
- TheoremedeLax-Milgram
- Lebesgue's theorems
- Lefschetz fixed point theorem
- Legendre Theorem
- Theorem Lehmann-Scheffe
- Leibniz Theorem
- Theorem Leray **(and)**
- Theorem Levinson
- Lévy's continuity modulus theorem **(in)**
- Lévy's continuity theorem
- Li-Yorke Theorem
- Lyapunov's central limit theorem
- Lie Theorem
- Lie-Kolchin Theorem
- Monotonic limit theorem
- Central limit theorem
- Lindeberg 's central limit theorem
- Lindemann-Weierstrass theorem
- Lindenbaum Theorem
- Linnik Theorem
- Liouville theorems
- Open book theory
- Löb's Theorem
- Lomonosov Theorem **(and)**
- Los Theorem
- Theorem Löwenheim-Skolem
- Lucas theorem
- Two-moon theorem

- Theorem Lüroth
- TheoremedeLusin
- Mackey's irreducibility theorem
- Theorem Mackey-Ahrens **(de)**
- Theorem Maclaurin
- Theorem Mahler
- Mahler 's compacity theorem **(of)**
- Malus Theorem
- Mann Theorem
- Marcinkiewicz interpolation theorem **(Theorem)**
- Theorem Marden
- Markov-Kakutani fixed point theorem
- TheoremedeMarkov-Post
- Convergence theory for martingales
- Theorem Maschke
- Theorem of Matiyasevich, or Robinson-Matiyasevich
- Maximum principle
- Maxwell Theorem
- Maxwell-Betti theorem of reciprocity
- Theorem Maxwell-Gouy
- Theorem McCoy
- Median theorem , or Apollonius theorem
- Mellin's inversion theorem **(Mellin's inversion theorem)**
- Theorem Menelaus
- Theorem Menger
- Theorem Mercer
- Theorem Mergelyan
- Theorem Mermin-Wagner-Hohenberg-Coleman
- Mertens Theorem
- Metsänkylä theorem

- Meusnier (or Meunier) theorem
- Meyer's Theorem
- Meyers-Serrin Theorem H=W
- Michael 's selection theorem
- Midy Theorem
- Midpoint theorem
- Miller Theorem
- Millman Theorem
- Milman-Pettis theorem
- Mills Theorem
- Milnor conjecture, demonstrated by Voevodsky
- Milnor conjecture (knot theory), demonstrated by Kronheimer and Mrowka
- Milnor decomposition theorem Milnor decomposition
- Milnor-Moore Theorem
- Minkowski Theorem
- Miquel's theorems (five circles , pivot, etc.)
- Mittag-Leffler theory
- Modularity theorem
- Mohr-Mascheroni Theorem
- Moivre-Laplace theorem
- Moment theorem
- Theorem of angular momentum
- Monge's theorem
- Monodromy theorem
- Montel Theorem
- Theorem Mordell-Weil
- Theorem Morera
- Theorem Morley (geometry)
- Morley 's category theorem
- Theorem Morley-Petersen

- Theorem Moser
- Theorem Motzkin
- Mustard Theorem
- Average theorem
- Cauchy mean theorem
- Mumford 's connectivity theorem
- ThéorèmedeMüntz-Szász

N

- Theorem Nagel
- Theorem Nagell-Lutz
- Theorem Nagumo
- Theorem Nakamura
- Napoleon's Theorem
- Nash Theorem
- Nash plunge theorem
- Theorem Nash-Moser
- Theorem Nernst
- Theorem Neuberg
- von Neumann theorems
- Theorem Newton
- Honeycomb theory
- Theorem Nielsen-Schreier
- Niemytzki-Tychonoff theorem
- Noether's theorems
- Strong law of large numbers
- Weak law of large numbers, or Khintchine's theorem
- Pentagonal number theorem
- Theorem of prime numbers, or Hadamard-de la Vallée Poussin theorem
- Norton's theorem
- Novikov's compact sheet theorem

- Nyquist-Shannon sampling theorem

O

- Optimisation/separation theorem
- Orlicz-Pettis theorem
- Osgood's theory
- Ostrowski theorem

P

- Painlevé-Morel theory
- Paley-Wiener theorem
- Butterfly theory
- Pappus Theorem
- Parthasarathy Theorem
- Pascal's Theorem
- Customs clearance theory
- Theorem of three perpendiculars
- Perron-Frobenius Theorem
- Peter Weyl Theorem
- Petersen Theorem
- Picard's theorems
- Picard'successive approximation theorem
- Pick's theorem Pick
- Theorem of Pierpont Theorem
- Theorem of Pitot
- Theorem of Plancherel Theorem
- Theorem of Pohlke Theorem
- Poincaré's conjecture, demonstrated by Perelman
- Poincaré's theorems
- Poincaré-Bendixson theorem
- Poincaré-Birkhoff theorem
- Poincaré-Birkhoff-Witt theorem

- Poincaré-Hopf theorem
- Fixed point theorems
- Five-point theorem
- Nine-point theorem
- Poisson's theorem
- Pólya theorems
- Poncelet theorem
- Poncelet's great theorem
- Poncelet-Steiner theorem
- Pontryagin duality theorem
- Door theorem
- Coat rack theorem
- Theorem of Post
- Theorem of Poynting
- Prigogine's theorem Prigogine's Theorem
- Projection theory
- TheoremedeProth
- Ptolemy's Theorem
- TheoremedePuiseux
- Theorem Pythagoras

Q

- Japanese theorem for inscribable quadrilaterals
- The momentum theorem
- Quillen-Suslin theorem
- The fifteen theorem

R

- Rademacher Theorem
- Rado's theorems
- Radon Theorem
- Radon theorem (geometry)

- Theorem Radon-Nikodym-Lebesgue
- Rajchman Theorem
- Ramon's Theorem
- Ramsey Theorem
- Rank theorem
- Constant rank theorem
- Theorem of rarefaction of prime numbers
- Lorentz reciprocity theorem
- Quadratic reciprocity law
- Recovery theory
- Reeh-Schlieder Theorem **(and)**
- Bearing theory
- Rellich-Kondrasov theorem
- Residue theorem
- Chinese remainder theorem, or Chinese theorem
- Crystallographic restriction theory
- De Rham's theorem
- Rice Theorem
- Richardson's Theorem
- Riemann theorems
- Riemann-Lebesgue theorem
- Riemann-Roch theorem
- Riesz theorems
- Riesz-Fischer theorem
- Riesz-Thorin Theorem
- Robbins Theorem
- Robertson-Seymour Theorem
- Rolle Theorem
- Rotational theory
- Rouché Theorem

- Rouché-Fontené theorem
- Routh Theorem
- Runge Theorem
- Rybczynski Theorem
- Fixed point theorem by Ryll-Nardzewski

S

- Theorem Salmon
- Sandwich theory
- Ham sandwich theory
- Sard Theorem
- Schauder fixed point theorem
- Scheme theorem
- Schmidt Theorem
- Schnirelmann Theorem
- Schoenflies Theorem
- Schreier refinement theorem
- Schur Theorem
- Schur-Horn Theorem
- Schur-Zassenhaus theorem
- Schützenberger factorisation theorem
- Schwartz kernel theorem **(of)**
- Theorem Schwarz
- Approximate selection theorem
- Separation theorems
- Root separation theorems
- Serre theorem
- Theorem Shannon-Hartley
- Theorem Shirshov-Witt **(and)**
- Theorem Shoute
- Theorem Siegel-Mahler

- Théorèmedusinge savant
- Theorem Simson
- Skorokhod representation theorem
- TheoremedeSleszyŋski-Pringsheim
- Theorem Slutsky
- Theorem Soddy
- Four-vertex theorem
- Spectral theorem
- Sperner's theorem
- Paradox of the sphere
- Theorem Sprague-Grundy
- TheoremedeStampacchia
- Theorem Stark-Heegner
- Staudt-Clausen theorem
- Theorem Stein
- Steiner's theorems
- Theorem Steiner-Lehmus
- Theorem Steinhaus
- Steinitz theorems
- Theorem Stewart
- Theorem Stickelberger
- Theorem Stieltjes
- TheoremedeStieltjes-Vitali-Montel
- Theorem Stokes
- Stokes' uniqueness theorem
- Stolz-Cesàro theorem
- Stone's theorems
- Stone-Weierstrass theorem
- Structure theorem for abelian groups of finite type
- Theorem Sturm

- Submersion theory
- Substitution theory
- Adjacent sequences theorem
- Sunyer i Balaguer theorem
- Supplementary orthogonal theorem
- Sylow theorems
- Sylvester theorems
- Theoremede Synge
- Theorem Szemerédi
- Théorèmede Szemerédi-Trotter

T

- Theorem Tabov
- Takagi existence theorem
- Theorem Tarski
- Theorem Tauber
- Theorem Taylor-Lagrange-Young
- Theorem Taylor-Proudman
- Theorem Chebyshev
- Theoremede Tellegen
- Theorem Terquem
- Théorèmede Thâbit ibn Qurra
- Thales Theorem
- Thales theorem (circle)
- Theorem Thébault
- Théorèmede Thévenin
- Thom 's isomorphism theorem
- Thom's Transversality Theorem
- Theorem Thue
- Theorem Thue-Siegel-Roth
- Thurston's conjecture, demonstrated by Perelman

- Tietze-Urysohn extension theorem
- Theorem Tijdeman
- Théorèmedutoit
- Theorem Torelli **(Theorem)**
- Theorem Torricelli, or of Schruttka
- Toyama Theorem
- Trace theorem
- Transfer theory
- Trapeze theorem
- Virtual work theorem
- Theorem of the triangle of forces
- Tsuji Theorem
- Turán Theorem
- Tverberg Theorem
- Tychonov's Theorem

U

- Uniqueness theorems

V

- Intermediate value theorem
- Theorem van Aubel
- Theorem van Cittert-Zernike **(and)**
- Théorèmedevan der Waerden
- Theorem van Kampen
- Varignon theorems
- Vaschy-Buckingham theorem, also known as the Pi theorem
- Vaught's theorem
- Viète's theorem
- Theorem Vinograd
- Theorem Vinogradov
- Viral theory

- Theoremede Viro-Sturmfels
- Vitali's recovery theory
- Vitali convergence theorem (complex analysis) **(from)**
- Theorem Lebesgue-Vitali
- Theorem Vitali-Hahn-Saks **(and)**
- Theorem Viviani
- Theorem Vizing
- Theorem Voronoi **(and)**

W

- Theorem Wallace-Bolyai-Gerwien
- Theorem Wantzel
- Theorem Wedderburn
- Weierstrass theorems
- Theorem Weierstrass-Casorati
- Weil's conjectures, demonstration completed by Deligne
- Theorem Weinstein
- Theorem Whitehead
- Theorem Wielandt
- Whitney's extension theorem
- Theorem Wick
- Wiener's Théorèmetaubérien
- Theorem Wiener-Khintchine
- Theorem Wigner
- Theorem Wigner-Eckart
- Wilson's theorem
- Witt's theorem
- Wittenbauer theorem
- Wolstenholme theorem
- Wronskian theorem
- Wulff's theorem

X

- Xian Tu theorem

Z

- Zariski's main theorem **(in)**
- The von Zeipel theorem
- Zermelo theorem

References

Aristotle, La métaphysique, Paris: Ladrange, (1838), (1840), 2008, p.35-36.

Hegel, G.W.F., Esthétique, Quebec, 1835.

Wikipedia ; logical link between theorem , available on What is the difference between theorem, reciprocal and contraposed? - Objectif : réussir en maths (objectif-reussir-en-maths.com) Accessed Thursday 2 November 2023 at 15 :48.

Wikipedia. 18 worksheets on the theorem of Thales of Milet, available on 18 worksheets on the theorem of Thales - Prof Innovant Accessed Thursday 2 November 2023 at 11 :26.

Wikipedia. Biography of Pythagoras, available on Pythagoras - Wikipedia (wikipedia.org) Accessed Thursday 2 November 2023 at 10 :37.

Wikipedia. Biography of Thales, available on Thales of Miletus: biography and philosophy (fílosofíadoinicio.com) Accessed Thursday 2 November 2023 at 11 :29).

Wikipedia. Contrapositive of Pythagorean theorem, available on Révision quatrième du théorème de Pythagore (free.fr) Accessed Tuesday 31 October 2023 at 8 :43.

Wikipedia. Definition of a theorem, available on Theorem: definition, applications and examples - Progresser-en-maths Accessed Thursday 2 November 2023 at 11 :00.

Wikipedia. Pythagorean theorem exercise, available on Pythagorean theorem and its reciprocal - Right-angled triangle (jaicompris.com) Accessed Thursday 2 November 2023 at 11 :45.

Wikipedia. Exercises on the Pythagorean theorem, available on Pythagorean theorem, course + corrected exercises. (paramaths.fr) Accessed Thursday 2 November 2023 at 11:15.

Wikipedia. Pythagorean theorem formula, available on Pythagorean Theorem : lessons and exercises | StudySmarter Accessed Thursday 2 November 2023 at 9 :45.

Wikipedia. La contraposée, available at contraposée | Lexique de mathématique (netmath.ca) Accessed Thursday, November 2, 2023 at 12:8.

Wikipedia. Reciprocity, available on Reciprocal involvement - Wikipedia (wikipedia.org) Accessed Thursday 2 November 2023 at 13 :59.

Wikipedia. List of theorems, available on List of theorems - Wikipedia (wikipedia.org) Accessed Thursday 2 November 2023 at 16 :15.

Wikipedia. Pythagorean theorem, available on Le théorème de Pythagore | Méthode Maths (methodemaths.fr) Accessed Monday 30 October 2023 at 16 : 13

Wikipedia. Pythagorean theorem, available on Le théorème de Pythagore | Méthode Maths (methodemaths.fr) Accessed Monday 30 October 2023 at 16 : 13.

Wikipedia. Theorem, available on Theorem - Wikipedia (wikipedia.org) Accessed Thursday 2 November 2023 at 10 :48.

Wikipedia. Theorem, available on Pythagorean theorem - Wikipedia (wikipedia.org) Accessed Tuesday 31 October 2023 at 9 :7.

Printed by Books on Demand GmbH, Norderstedt / Germany